THE REFERENCE GAME

A USER'S GUIDE TO THE FIRST 2 DIMENSIONS OF

INFINITY COURT

DANIEL A. MARTENS YAVERBAUM

Infinity Court Records
From the Inner Shell Delegation
to 3rd Loop, Axis XVIII,
the **United** *F* of ***R***

printed here with permission and support
by its duly affiliated terrestrial imprint at

Law Four Press
Brooklyn, NY 11213

COPYRIGHT

ISBN: 9780998584768

CONTENTS

PART I
ROUND ZERO

1

THE PLAYERS

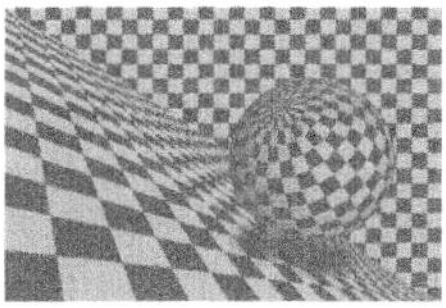

PLAYER ONE: DATA MASTER

Alias:

- *Dealer.*

ALSO:

- *Thedama,*
- *Haus,*
- *Prime Mover,*
- *Croupier d' Crappes Cosmologique,*
- *Gott,*
- *Dadamastah.*

ESSENTIAL ROLE:

set initial conditions;

administer measurements.

AFFILIATIONS, ASSOCIATIONS (IF APPLICABLE):

the *witness**ed***—

those found or designated to be on the *receiving* end of some

study, inspection, examination, observation, measurement, look, gaze, watch, tick, click, ping, mark, tally or similar interval of attention.

RECOMMENDATION—REGARDING THE ASSIGNMENT of Roles:

- For best standard play of *Round 0*, it is strongly recommended that the role of *Data Master* be assigned to the more experienced of the two players.
- *Round 0* is uniquely designed in order to induct a first-time player into the game expanse; the *Data Master (*from here on, 'Thedama'*)* directs the initiating bout from within the competitive ring — functioning at once as opponent and guide.

COMMENT—REGARDING THE (ABOVE) Recommendation:

- The assignment of Thedama's role to the *more* experienced of two players is, as has been stated, strongly recommended in typical round settings.*
- It is, of course, quite possible to be the more experienced of two and yet be hardly experienced at all. Taken literally, the above recommendation does not seem to exclude any particular candidate nor class from Thedama's seat.
- This is neither misunderstanding nor accident. A good Thedama might well lack largely or entirely in experience. In the latter case, presumably, the same would be true for Thedama's opponent. A match-up of this kind offers its own distinct pallet of potential diversions and rewards. Such game-play is, to be sure, fully supported.
- A large group or classroom of pairs, for example, can be broken into pairs and overseen by a teacher or professor as they conduct their own initiating rounds.
- In sum, two first-time players can discover and navigate the Game straight out of the box. This exciting possibility is facilitated by two distinct instruction documents, each privately addressed to one of the two player roles. The packet is included with all Game materials and reprinted with annotation here (see *Launch Letters*).

*See *Notes on Round Zero #1: Ex Nihilo Nihil*

PLAYER TWO: OBSERVER

Alias:

- *Sentinel.*

Also:

- *Aneyewit,*
- *Surveyer,*
- *Scholar,*
- *Enumerator,*
- *Technician,*
- *Registrar,*
- *Watson,*
- *Playah.*

ESSENTIAL ROLE:

make measurements;

find particles.

AFFILIATIONS, ASSOCIATIONS (IF APPLICABLE):

the *witnesses*—

those found or designated to be on the *viewing* end of some

study, inspection, examination, observation, measurement, look, gaze, watch, tick, click, ping, mark, tally or similar interval of attention.

2

THE BOX

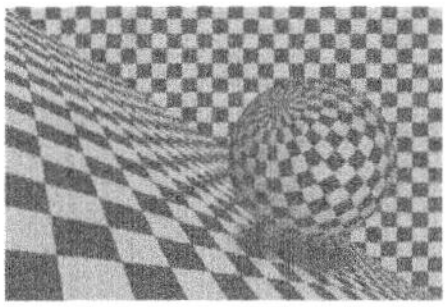

THE LAUNCH LETTERS

- In order to guarantee universal access to a sound and basic *Game* education, the *Round 0* box includes two distinct *'Launch Letters'* of greeting and direction, each addressed exclusively to one of the two players.
- Immediately upon assignment of roles, each player must open and read his/her personal *Launch Letter*. Care must taken to maintain strict confidentiality.
- By means of such *Letters*, an independent and inaugural *Round 0* can be played — with a freshly opened box and without need for third party nor external resource — by two first-time players.

THE BOARD

Alias:

- *Grid.*

Also:

- *Coordinate System,*
- *Reference Frame,*
- *World.*

DESIGN:

The game board is a thick *square matrix*:

- 10 identical *rows*, by
- 10 identical *columns* comprise,
- in total, 100 identical *cells*.

The shape of the matrix itself exhibits *90-degree rotational symmetry.* In preservation of this geometric property:

- no side nor portion of the board is labeled,
- all adornments are uniform or strictly patterned;
- the four sides of the board remain indistinguishable from one another.

CELL STRUCTURE:

- Each *cell* is a physical indentation, groove, aperture or similar concavity carved from the otherwise smooth surface of the board.
- Each *cell* is just large enough to contain precisely one '*piece*' (see *The Pieces*, below).
- Each *cell* is equipped with an opaque shutter, valve, screen, cover or similarly movable flap.

CELL IDENTIFICATION:

- For provisional application to a single round of play, one side of the square is designated the *top* of the matrix.
- The designation is arbitrary, but must be known to both players.
- This designation is among the first things to occur in the round. It is made by Thedama prior to any interaction with Aneyewit (see *The Set-Up*).
- Thedama is expected to share his designation with Aneyewit immediately upon Aneyewit's entry to the *Game-Frame*.
- Once the two players meet each other, the board is oriented so that the stipulated side is nearest to Thedama. The side will thereby appear and serve as *top* from the *vantage point*** associated with Aneyewit's seat. The side opposite *top* becomes *bottom*.
- Cell *rows* are to be kept unmarked. Counting

from *top* to *bottom,* the rows are nonetheless identified as *Row 1* through *Row 10.*

- Facing the *top* and counting from the left to right, cell *columns* are labeled *Column 1* through *Column 10.*
- Each cell is thereby uniquely identified by a coordinate pair, presented in (*Row, Column*) form [from here on, (*R, C*) form].

**See *Notes… #2: Vantage Point*

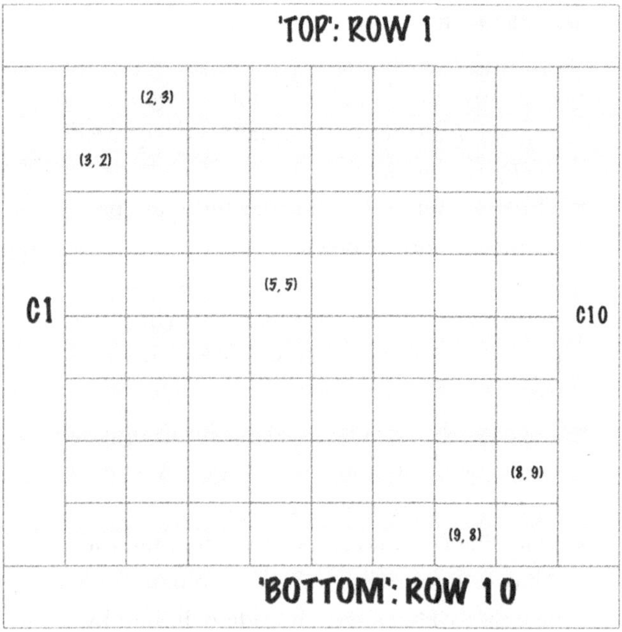

A Properly Configured Game Board

CELL STATES:

At any given moment of play, any given cell is either

1. *occupied* (by a piece) or
2. not occupied (thus empty);

At the same moment, moreover, the same cell is either

1. open (for measurement) or
2. not open (at default).

In order to start a new round of play, every cell must be

1. not occupied and
2. not open.

When and only when a cell is selected for measurement, then

1. Thedama opens the cell *shutter*,
2. Thedama notes the contents of the cell interior,
3. Thedama closes the cell *shutter*.
4. Thedama, at some point prior to the end of his turn, somehow shares and/or reports the result of measurement to Aneyewit.
5. Thedama confirms the cell shutter to be secure in its normal closed state.

THE PIECES

Alias:

- *Tokens.*

Also:

- *Objects,*
- *Subjects,*
- *Actors,*
- *Agents,*
- *Amplitudes,*
- *Eigens,*
- *Entities,*
- *Occupants.*
- *Things.*

POPULATION:

The box includes a full *population* of pieces, consisting of

- a total of fifty (50) small solid discs,
- each composed of ceramic or similarly durable material,
- each of identical size (approximately that of a coat button):
- a single piece can occupy a cell;
- a single cell can contain a piece.

COLOR—ATTRIBUTABLE TO *PIECES*

- Pieces can be distinguished from one another only sometimes and only with reference to color.
- The full set of pieces can be sorted into 5 groups, each characterized by a distinct color.
- Each color group contains 10 identical pieces.

PIECES V. PARTICLES

Piece: Essential Character

A *piece*, on the one hand, is

- an actual tangible token (manipulative),
- physically present and deployed during game play,
- in order to represent, indicate, symbolize or
- model, in part or in whole,
- the presence in principle of a *particle*.

PARTICLE: ESSENTIAL CHARACTER

A *particle*, on the other hand, is

- a single, discrete, identifiable element,
- subject to great mental concern and attention throughout game play,

- yet ever inaccessible by any direct means to the bodily senses;
- an archetype for the entire known population of tokens, objects, subjects, actors, agents, eigens, entities, occupants, or like things
- which can inhabit or traverse a coordinate system, matrix board or world.

COLOR:

- *IS* attributable to *pieces*,
- *IS NOT* attributable to *particles*

At increasingly advanced rounds of play, the variable attribute *piece color* can support an increasingly wide array of implications and uses. The beginning stages of play, however, are of primary concern here. In early rounds and with inexperienced players, it seems, piece color presents a potential source of confusion or distraction. Directly below, therefore, the topic of color as it pertains to *particles* is explicitly contrast with that of *pieces*.

- Unlike pieces, no single particle can be visually identified nor otherwise characterized by any particular color.
- Unlike pieces, there is no common trait nor characteristic with respect to which one arbitrarily selected particle can be readily distinguished from another.

ONE-TO-ONE CORRESPONDENCE:

- The number of pieces concealed in cells need NOT identical to the number of particles represented.
- Occupation of a cell by a piece indicates the occupation of the world by a particle, but the indication need not necessarily be one-to-one.
- Depending on round and context, for example, a full column of physical pieces is commonly used in order to represent a single particle.
- In no case can the relation between the sizes of the two populations be arbitrary. There must be some sort of coherent mapping from one to the other.
- In general, the particles are the items of interest; the pieces are vivid symbols, close approximations, physical pawns, analytical tools used to detect particles.

3

THE GAME-FRAME

Alias:

- *Frame.*

Also:

- *the Frame,*
- *Frame of Reference,*
- *Perspective,*
- *Arena,*
- *Velocity,*
- *View.*

DEFINITIONS

Preliminary Denotations:

1. the region of spacetime occupied by both players, by the board, by the pieces, and by all other items and materials used in a round of play (see *The Box*, above, and *The Game Play*, below);
2. the single uniform velocity maintained continuously and characteristically by both players along with the board and all associated materials (from here on, '*the board, etc.*') for the round's entirety;
3. the commonality of velocity by virtue of which each player necessarily remains stationary in relation to her opponent, to the board, etc.

Larger Implication:

Of perhaps most pertinent yet most perilous meaning is the following use of the term *Game-Frame*:

4. The perspective or world-view
uniquely determined by the particular velocity
attributable to the Frame
and maintained identically by the players
occupying the Frame.

Some elaboration is warranted.***

- Any quantitative finding reached by either player in reference to any motion external to the *Game-Frame* is tethered to this shared perspective.
- Of any spacetime interval external to the *Game-Frame*, measurement from the *Game-*

Frame is accurate only provisionally or conditionally: The measured quantity is accurate only to the extent that the *Game-Frame* perspective is applicable.

- No matter how distant, anything and everything traveling at a velocity identical to that of the *Game-Frame* maintains a perspective identical to that the *Game-Frame*; all that travel with identical speeds and directions will obtain identical results from measurement.
- No matter how proximate, anything and everything traveling at a velocity different from that of the *Game-Frame* will maintain a perspective different from that of the *Game-Frame*; all that travel at another speed or direction will obtain different results from measurement.
- For heightened emphasis on the fourth and somewhat figurative meaning of the term *Game-Frame* use of the alias *Frame of Reference* is recommended.

***See *Notes… #3: Game-Frame - Larger Implication*

INTERIOR CONTENTS

Game Table:

Within the *Game-Frame* is a *Game Table,*

of area sufficient to accommodate comfortably the *Board* in addition to all the items listed below:

- two stop-clocks set at opposite corners of the table,
- two identical cocktail bowls and
- two identical cocktail glasses, set at opposite corners of the table.***

*** See *Notes... #4: Game Frame - Some Assembly Required.*

~

Game Referees:

All items listed above are to be managed vigilantly throughout game play by two referees (or soldiers, special operatives,

- credentialed by authorized agency of the infinity-court,
- selected at random,
- liveried,
- unarmed,
- remaining at attention
- located in arm-length diagonals from the two unoccupied corners of the Board.

~

Hover-Floor:

One sturdy circular (cylindrical) floor, platform, stage, dais or similar, sized and situated as customary floor, yet also:

- refurbished with *hover-form* mount & housing,

- securely bearing *'preference frame'*-grade hover-board engine,
- mount, housing and engine certified as inspected by the *Commission for Safety of the Infinity Court,*
- engine customized to game-scale,
- engine approved for stability up through and including geomagnetic storm magnitudes,
- certificates prominently displayed.

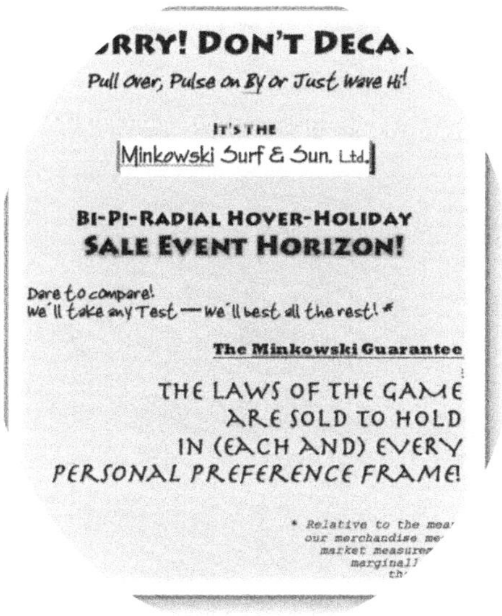

Court-Approved Manufacturer of Preference Frames

4

THE SET-UP

Alias:

- *Shuffle.*

ALSO:

- *Big Bang,*
- *Genesis,*
- t_0.

ENTER THEDAMA

Thedama is to enter the *Game-Frame* first.

S/he is to do so unaccompanied.

By means of the following steps,

Thedama sets the initial conditions.

1. Thedama must first read or at least ready the *Launch Letter* addressed to her. Even if familiar with its contents, she must be prepared to deliver verbatim to Aneyewit a script contained within.
2. Thedama must then select a sample of pieces — whatever number is needed to represent 1 - 3 particles.
3. Thedama is to place each one of the chosen particles into any one unoccupied matrix *cell* of her choosing -- one per one.

Thedama must make careful note of:

- the (*R*, *C*) coordinate pair for each occupied cell &
- the color of each cell occupant.

When and only when fully satisfied both with the accuracy of her notes and with her command of the *Launch Letter,* Thedama is to invite the Aneyewit into the *Game-Frame.*

- Taking care to withhold any/all indications of particle placement, Thedama must deliver the script found within the Launch Letter.
- Script must be delivered verbatim.
- As per script, Thedama will inform Aneyewit that '1 - 3' particles are represented by pieces dwelling within the matrix cells, 'one per one'.
- As per script, Thedama will inform Aneyewit of game object (see *The Objective,* below).

5

THE OBJECTIVE

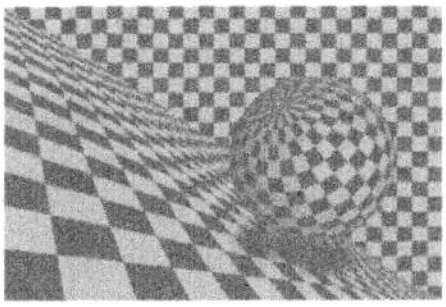

Alias:

- Goal.

ALSO:

- *Purpose,*
- *Telos,*
- *Destiny,*
- *Research Motivation,*
- *Scorecard*

STATEMENT THEREOF

Aneyewit's Goal:

In order to win, Aneyewit must *identify* the full sample of particles selected by Thedama before the declaration of *Game's End.*

An adequate identification must account for:

- all *(R, C) coordinates* and *colors*
- necessary and sufficient
- to distinguish the full *sample* of *particles*
- from the general *population of particles* and
- from **one another**.

THEDAMA'S GOAL:

In order to win, Thedama must prevent Aneyewit from completing identification of the particle sample, as specified above.

THE CONSTRAINT: GAME'S END

- *Game's End* arises at the first moment a certain predefined set of conditions are met; the moment this '*Game's End Function*' is satisfied, Thedama must announce as much. No more play follows.
- The '*Function*' which determines the arrival of the Game's End is not arbitrary. It can be be

stated and tested independently of the circumstances for any particular round.
- The two above facts appertaining to the Game's End function are central to the play of Round Zero. No measurement can be made before Aneyewit is made explicitly aware of both facts. For this purpose, Thedama's letter includes a script. During *Set-Up,* Thedama must deliver this script **verbatim** to Aneyewit (see Chapter *7, Launch Letter the First*).

THE METHOD: MEASUREMENT

In order to find particles, cells are measured, one measurement at a time.

From the report provided by Thedama regarding any given measurement, Aneyewit can learn one of two things:

1. We observe a particle in cell (R, C) OR

2. We do not observe a particle in cell (R, C).

SCOPE:

- The spatial scope of a given measurement does not extend beyond the cell boundaries.
- Similarly, the temporal scope of a given measurement should not be taken as a permanent statement.
- The cell data are neither arbitrary nor

misrepresented, but they can and do change over the real-time course of a round.

- Two successive measurements of the same cell **can** and often **do** yield two different results.
- Differing measurement results for the same cell can occur in any round, at any level of play.
- Differing reports for a single cell ought not be dismissed as random error, nor should they be feared as indicative of a fundamental confusion.
- In higher rounds, the possibility barely invites explanation: particles can move.
- Here in *Round Zero*, particles do not move: Some other explanation must apply to the possibility of differing measurements for the same cell.

Multiple measurements of a single cell can differ from one another on the following basis:

- Every report makes reference to what is or is not observed.
- A report of such kind consists of relational information.
- It might be intended to describe a single object, but it potentially contains just as much information for another.
- In his *Launch Letter*, Thedama is explicitly instructed that his response to each measurement will depend on what he observes in the selected cell.
- It will also, however, depend on certain

features of the observation process (selection sequence) itself. The report will partially depend on an *'extra-cell'* factor which is present and observable whether or not the cell shutter is ever opened.

THE CONSIDERATION: EXTRA-CELL REPORTING

- The precise function of an extra—cell feature is unambiguous and potentially auto-executable.
- The details of this function **must not** be shared with the Aneyewit; likewise, the fact of its existence **must** be shared.
- With respect to the above matters of confidentiality and objectivity, the function governing *Extra-Cell Reporting* operates throughout game play in much the same manner as does the function determining *Game's End*.
- The confidential details of both functions are provided in Thedama's *Launch Letter*.
- The confidentiality and objectivity associated with the two functions must be explicated to Aneyewit by means of verbatim delivery of the script found in Thedama's *Launch Letter*.

6

THE GAME PLAY

(THE MEASUREMENTS)

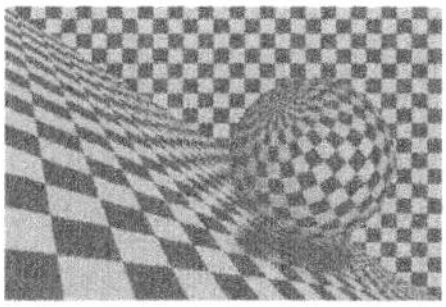

TURNS

First, Aneyewit's turn.

- Aneyewit is allotted precisely 90 seconds to make his measurement, accomplish any/all subtleties and complete his turn.
- Strict Timekeeping is enforced by the referees.

Second, Thedama's turn.

- the Thedama is allotted precisely 90 seconds to accomplish a wider possible array of less sharply defined tasks.
- Strict Timekeeping is enforced by the referees.

Turns will continue to alternate in the above fashion until Game's End.

The contents of any given turn are specified below.

NOTE—No Time Dilation:

For the full duration of a round,

- The two players will remain in a single *Game Frame ('of Reference')*
- and thereby at rest relative to each other.
- With reference to the time interval between any pair of events, therefore,
- both players should always obtain identical measurements.

MEASUREMENT

Definition:

The full act of measurement and associated protocol comprises:

1. direct observation by Thedama of a cell interior,
2. consultation by Thedama of notes from all past measurements,
3. consultation by Thedama of any/all Law, if applicable.

ANEYEWIT

The opening half of each turn belongs to Aneyewit.

- Once Aneyewit's timer begins, his job is to select for measurement one of the 100 possible matrix cells.
- The selection is unconstrained and entirely his choice.
- The whole of his time allotment is available for cell selection.
- The whole of his power and challenge through the round is that of cell selection.
- Before the time elapses, Aneyewit must inform Thedama of his cell selection.
- It is incumbent upon Aneyewit to keep careful notes/records of all his cell selections.

THEDAMA

The closing half of the turn belongs to Thedama.

- Thedama will likely begin by opening the shutter belonging to the specified cell and examining the interior.
- This direct observation need not be made visible to Aneyewit.
- (Some manner of disclosure is obligatory, but, for reasons to become evident below, the

precise manner of disclosure depends on the precise details of the cell/shutter design.)

- Thedama will then consult his own personal notes/record of the round thus far.
- Thedama is then encouraged to refresh his recollection and application of the Law by means of reference to this volume and/or similar texts and tracts.
- For much of the remaining time period, Thedama might engage in silent and intense contemplation or he might deliberately give the false impression of doing so.
- He might address the Aneyewit on any number of topics;
- a topic might bear obvious and direct application to the measurement at hand; it might swing wildly between acute deposition and desperate fishing expedition.
- Nothing said by Thedama obligates Aneyewit, in any statutory or enforceable manner, to say nor do anything in response.
- Nothing in the Law, moreover, prohibits any particular verbal (or otherwise non-violent) response on the Aneyewit's part.
- Thedama is under no obligation to share the result any earlier than the last possible opportunity. There is no spectrum of incentives nor motivations designed to honor anything past a bare minimum.
- No matter who says or does what when, it is fundamentally Thedama's time. She has the last word on the last word. Until that point, the Aneyewit must simply endure.

Any and all interchanges during Thedama's turn are limited and/or driven by three facts, both known to both players:

1. Thedama is in possession of the measurement result (and does not particularly need it).
2. Aneyewit needs the measurement result (and does not have it).
3. Thedama is obligated to share the measurement result with Aneyewit at some point during his turn – before the clock stops.

THE MEASUREMENT REPORT

Thedama will provide one of two possible reports:

1. We observe a particle in cell (*R, C*) *OR*
2. We do **not** observe a particle in cell (*R, C*).

CONTINUATION OF TURNS

All subsequent measurement turns proceed identically.

Measurement continues until declaration, as per script, of *Game's End*.

REMATCHES & REPETITION OF ROUNDS

It is recommended without reservation that Aneyewit play and replay Round 0 – with the same or various Data Masters – until he has come to understand the theoretical framework and achieved the objective or vice-versa.

SIMILARLY, IT IS RECOMMENDED WITHOUT reservation that Thedama refrain from divulging facts or functions until the above has occurred. Progressive series of hints, Socratic dialogues and similarly scaffolded supports are acceptable and encouraged.

7
LAUNCH LETTER, THE FIRST

TO THE DATA MASTER

Dear Player:

On behalf of the Infinity Court and Greater Metro Finity Court Sectors,

Welcome! Welcome!

Welcome, in particular, to the Zeroth Round for the 2-Player, 2-Dimension Edition of *The Game*.

IN THIS ROUND, YOUR ROLE IS THAT OF THE *DATA Master*. You are to be congratulated for achieving this honor.

- You will begin the game round in solitude and set up the round in secrecy.
- First, you will choose 1 - 3 particles for service as initial conditions of a world.

- In order to represent such particles, you will then select the appropriate number of game pieces from the full population (provided in game box).
- You will then place each one of them in a board matrix cell of your choosing.
- You will make careful note of the locations and colors of all the matrix inhabitants.
- Then and only then will you invite into the *Game-Frame* your opponent, the *Observer*.

~

SCRIPT FOR THE OBSERVER

To your opponent, the *Observer*, you will summarize the game set-up and objective precisely as follows:

"I am the *Data Master*."

- From a far larger possible population, I have selected a sample of 1 - 3 particles for habitation of the *game world*, as represented by this *board matrix*.
- In order to represent this sample of particles, I have placed *pieces* in *cells* of the matrix, one per one.
- NOTE: Pieces are used to represent particles, but representations can be subtle. The correspondence between pieces and particles ***need not*** be ***one-to-one***.
- Your objective is to identify the full sample of particles prior to *game's end*.
- I will declare *game's end* the moment it occurs.

My declaration might come quite unexpectedly, but it will not come arbitrarily.
- *Game's end* will occur for very specific reason – as a result of a very specific combination of factors and/or events. This specific combination is unambiguous and holds independently of any game circumstances. I could, in fact, fully express the function right now. But I will not."

~

TO YOUR OPPONENT, THE *OBSERVER*, YOU WILL THEN explain game play and measurement as follows:

- "In order to pursue this objective, you will choose one cell per turn for measurement.
- Each turn, you will provide me with the (*R*, *C*) coordinates corresponding to the cell you've chosen.
- I will perform the measurement and provide you with one of two reports, either:

1. 'We observe a particle in the cell' OR
2. 'We do not observe a particle in the cell'."

~

TO YOUR OPPONENT THE *OBSERVER*, YOU WILL continue:

"The content of my report will depend strictly and unambiguously on 2 factors:

1. direct observation of the cell interior: whether or not I do, in fact, observe a particle there;
2. a fully specifiable feature of your selection and/or prior selections: an 'extra-cell' function."

- "A precise set of rules (a function) dictates the manner in which I incorporate these factors into my report.
- I will not stray from this set of rules.
- Yet, I will not disclose them to you.
- In these respects, the extra-cell function operates much like the *Game's End* function.
- Great fortune to you, my worthy opponent and keen observer."

WHAT FOLLOWS, *DATA MASTER*, ARE THE PRIVATE facts and functions appertaining to Round 0 and to which your opponent, the *Observer*, must have no access:

THE RECOMMENDED ARRANGEMENT OF PIECES

- Choose to place precisely 1 distinct particle into the world
- In order to represent this particle:
- Place 10 pieces of a single color into the cells

spanning one full column, top to bottom of the matrix.

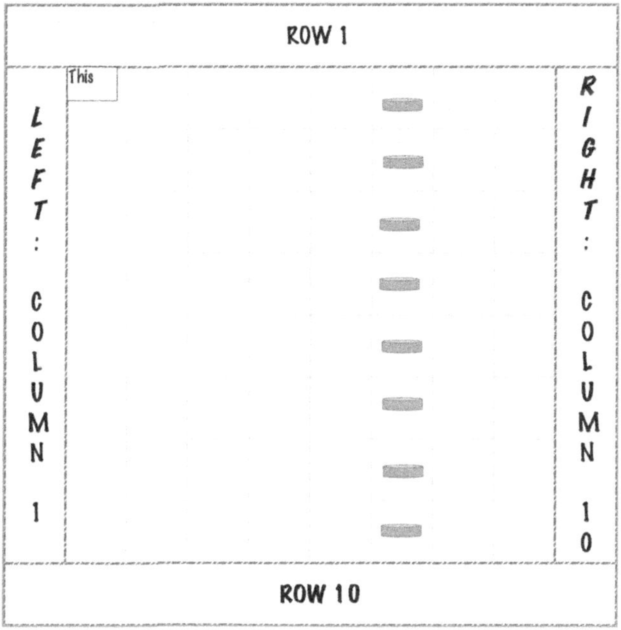

Recommended Placement of Pieces for Round Zero

(How to Report Measurement Results)

- Let the coordinates for the cell selected for this turn be (R, C).
- Unless this is the first turn of the round, let the coordinates for the cell selected for the turn directly prior to this one be (r, c).
- If this is the first turn, then let $(r, c) = (0, 0)$.

Given the above,

'We observe a particle in the cell' should be reported

if and only if both of the following conditions are met:

1. There is, in fact, a piece in the selected cell and
2. $r < R$.

Under all other conditions, 'We do not observe...' must be reported.

THE GAME'S END FUNCTION

Game's End occurs once the 10th row is selected for measurement.

Immediately upon the selection of the 10th row:

1. Thedama must provide the measurement result.
2. Thedama must declare the *Game's End*. Accordingly,
3. Aneyewit has the opportunity to propose his best possible identification of particles, given all information to that point.
4. Fourth, the Game is Over.

MODEL/THEORETICAL FRAMEWORK

- For this and all rounds to follow,
- each column represents a position in space;
- *each row represents an instant in time.*
- No self-evidently intrinsic nor directly observable property distinguishes one pure (unoccupied) dimension from the other.
- Once things start occupying dimensions, and once we start examining those things, then it starts seeming reasonable to distinguish one dimension (space) from another (time). Our collection of data seems constrained along one axis while not along the other.
- Regarding data collection, that is, both axes seem to be *homogenous* while only one seems to be *isotropic*.
- The function governing observation reflects our inability to collect data from two measurements made with reference to anything other than two successive instants of time.
- A full column of cells, for example, represents a single, stationary and wholly unremarkable particle: It holds steady in its occupation of one given point in space, while remaining observable and observed through an extended continuous sequence of instants in time.

✻✻✻✻✻ SEE *NOTES...* #5: *COLUMN CONSTRAINTS & Null Knowledge*

8
LAUNCH LETTER, THE SECOND

TO THE OBSERVER

Dear Player:

On behalf of the Infinity Court and Greater Metro Finity Court Sectors,

Welcome! Welcome!

Welcome, in particular, to the Zeroth Round for the 2-Player, 2-Dimension Edition of *The Game*.

IN THIS ROUND, YOUR ROLE IS THAT OF THE *Observer*. You are to be praised for your courage and and enthusiasm.

Your opponent is the *Data Master*. Your fierce adversary, this very same *Data Master* remains a most dedicated mentor and role model.

The remaining instruction, guidance and challenge

will come from the strange and wondrous maestro that is your *Data Master*.

Heed then well how the *Data Master* moves; heed well what s/he says.

Game's Beginning comes presently.

We wish you the smoothest and most continuous of motion.

9

NOTES ON ROUND ZERO

* #1. EX NIHILO NIHIL

Round Zero is the one and only round in which decisive information is withheld from one player by anything or anybody other than the opponent.

It is the only round in which information is withheld prior to and independent of game play — the only round which relies for its drama on a limitation of knowledge.

The practice is just sufficiently suspect to warrant comment, yet, according to word from the bench at Infinity Court, just sufficiently rousing to warrant a single implementation.

~

This treatment of information is tailored to

the singular pedagogic demands imposed by the very first lesson in any discipline.

- Like a very first lesson, we here endeavor to expose meaning in the absence of reference.
- We mimic an attempt to grow something from nothing.
- We inadequately imitate the boundary conditions for a small kinematic experiment or a big bang.
- We seek to model, in the end, the experience of trying to manage a class full of people all of whom know you forgot to assign homework.
- It is a concession: Never again in this game will the possibility of knowledge dare threaten to 'spoil' anything.

** #2. VANTAGE POINT

A vantage point must not be confused with a reference frame. The former is a position (point in space) which gives rise to a visual perspective. The scope and application of such perspective is essentially limited to this context — board placement.

The latter is a velocity (speed and direction through space) which gives rise to a measurement perspective. The meaning, scope and application of such perspective cannot be emphasized enough. For preliminary discussion, see *The Game-Frame*, above and below.

*** #3.GAME-FRAME - LARGER IMPLICATION

At early stages of play, this fourth meaning can seem problematic in any number of ways: The reference can seem needlessly figurative, inappropriately symbolic, deceptively self-evident or alarmingly insensible. It can, furthermore, seem innocuous precisely to by whom the concept is subjected to the most confounding of misuses.

For optimum appreciation and treatment of the Game-Frame at its deepest level of implication, there is but one central recommendation: extensive practice.

The truly seasoned player is an expert navigator. She feels unflappably at home in two particular regions of spacetime: that specific zone of spacetime enclosed within the Game-Frame and that surrounding vastness of spacetime found anywhere outside the Game-Frame.

**** #4. GAME-FRAME - SOME ASSEMBLY REQUIRED

The Game-Frame is indispensable to all that is intended, accomplished, and symbolized by game play.

It is possible and, in fact, overwhelmingly common

for the Game-Frame to perform the full sweep of its job description in a symbolic or figurative manner.

The presence of guards in livery, cocktail goblets, near-luminous speeds, etc., are absolutely guaranteed to optimize game play.

They are not, however, included in the box. Some assembly may be required.

***** #5. COLUMN CONSTRAINTS & NULL KNOWLEDGE

- Evidence for an axis (time) which distinguishes itself from the others (space) is found in the relation between data and its collector rises in The relation between collector and data (observed and observer). regarding a particular axis, and is not demonstrated otherwise.
- To signify an asymmetry intrinsic to the dimensions themselves, independent knowledge regarding the number and nature of occupants must first be supported or assumed.
- For example, we can assert and demonstrate with confidence that a chair can occupy one space at two times, while it cannot occupy two spaces at one time.
- But we can easily observe a pair of chairs audition for both scenarios. It's just that in order to accept one, we must have previously

decided to call the two identical-looking items at t_1 and t_2 the same chair.

- To reject the other, we have previously decided to call the two identical-looking items at x_1 and x_2 two different chairs.

PART II
ROUNDS ONE THROUGH FOUR

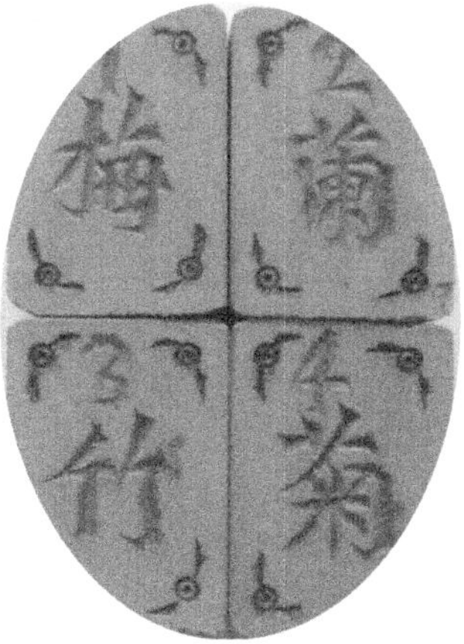

1

ROUND ONE

Play is identical to that of *Round Zero*, but for the following:

- At the beginning of Set-Up, prior to privately placing pieces, Thedama will select one of four possible directions along the board as the *time axis*.
- Upon being joined in Set-Up by Aneyewit, Thedama will demonstratively but carefully rotate the entire the board a few times as though to shuffle or scramble the axes.
- If desired, Thedama might even make a point of switching and/or shifting the two player chairs.

Thedama will then assure Aneyewit that in most respects the Round One objectives, procedures and board composition are identical to those of Round Zero. For example, he will explicate:

1. The board continues to consists of one spatial and one temporal axis.
2. It continues to be Aneyewit's goal to identify the full particle sample before Game's End.
3. As turned out to be the case with Round Zero, there is but one particle present in the board.

BUT:

- There is no reason to assume that the specific alignment of the board's temporal axis depends in any way on the circumstantial arrangement of items within the Game - Frame. Hence,
- The direction of the temporal axis is ambiguous; it is one of four possibilities.
- In order to pursue any information needed for the identification of particles, the means available to Aneyewit remains: measurement.
- Game's End remains Game's End, as per Round Zero.

RECOMMENDATION FOR THEDAMA:

Again, place a full column of pieces for the representation of but one particle in the manner of *Launch Letter, the First*.

This time, however, make sure the column's orientation is arbitrary compared to its orientation for *Round Zero*. This round, that is, Aneyewit will again search for 10 pieces — each of which represents the same one particle at the same one position of space, yet at a different instant of time. To his vantage point this round, however, the arrangement might look like a row. And it might even be a row that 'flows' right to left rather than the other way round. Pick any of four possibilities, such as that depicted below.

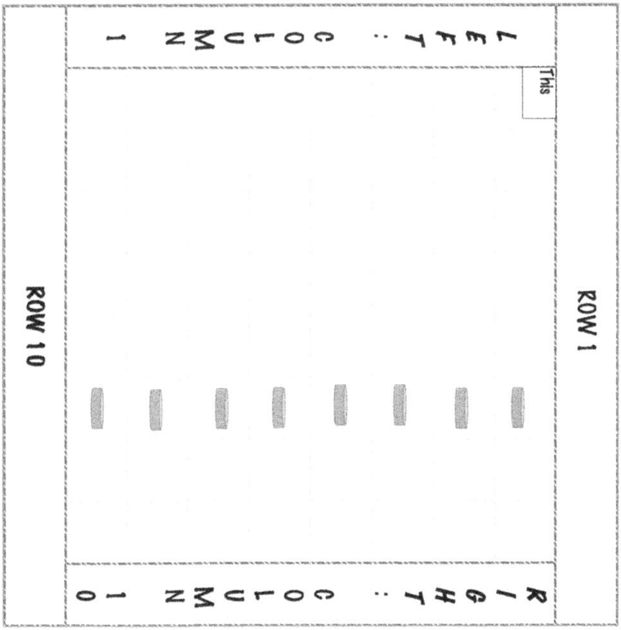

Recommended Placement of Pieces for Round One: Unknown Orientation of Time Axis, One Stationary Particle

NOTE: THERE IS NOW, AND FROM NOW ON, NO extra structural deception. The mystery to be solved is precisely as described. And the solution is entirely at Thedama's disposal.

2

ROUND TWO

Play is identical to that of *Round One*, but for the following small adjustment:

- As Thedama will tell Aneyewit,
- This time, *1 - 3* particles have been placed.
- All must be identified.

RECOMMENDATION FOR THEDAMA:

Again, place three full column of pieces for the representation of three one particles in the manner of *Launch Letter, the First*.

Make sure the orientation of the columns orientation is arbitrary compared to that of *Round One*.

For example,

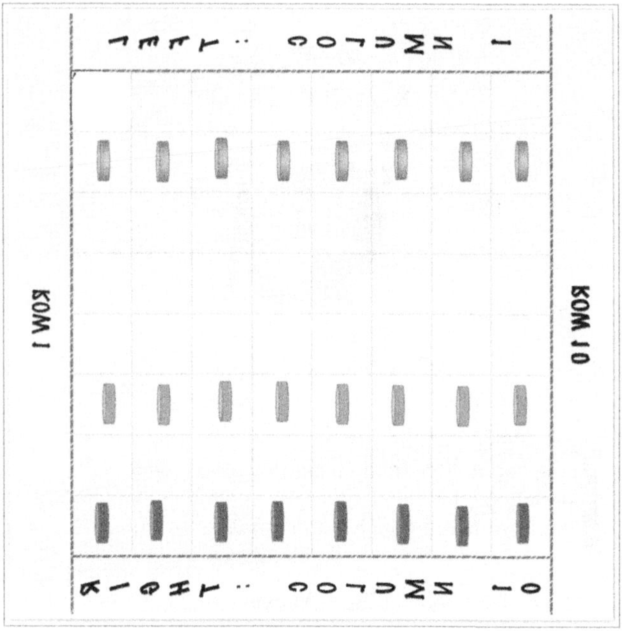

Recommended Placement of Pieces for Round Two: Unknown Orientation of Time Axis, Three Stationary Particles

~

3

ROUND THREE

Play is identical to that of *Round Two**, but for the following small adjustment:

As Thedama will tell Aneyewit,

- For this round, the sample size has shifted back to 1 - 3 particles.
- From here on, the use of static position coordinates will not suffice to identify a particle;
- Every particle found in a previous round happened to occupy a fixed point in space, i.e.: remain stationary.
- There is no reason to assume permanence or universality of whatever fortunate circumstances allowed for such a special and convenient arrangement.

**The board continues*

1. *to include a temporal axis,*
2. *to align the temporal axis (as well as the remaining system) along one of four possible orientations,*
3. *to be oriented in according to a choice made by Thedama, and withheld from Aneyewit.*

These three characteristics of the coordinate system represented by the board are assumed of all (increasingly advanced) rounds to follow until/until otherwise specified.

RECOMMENDATION FOR THEDAMA:

Again, select but one particle — one that moves at constant velocity through the board space.

CONSIDER, FOR EXAMPLE:

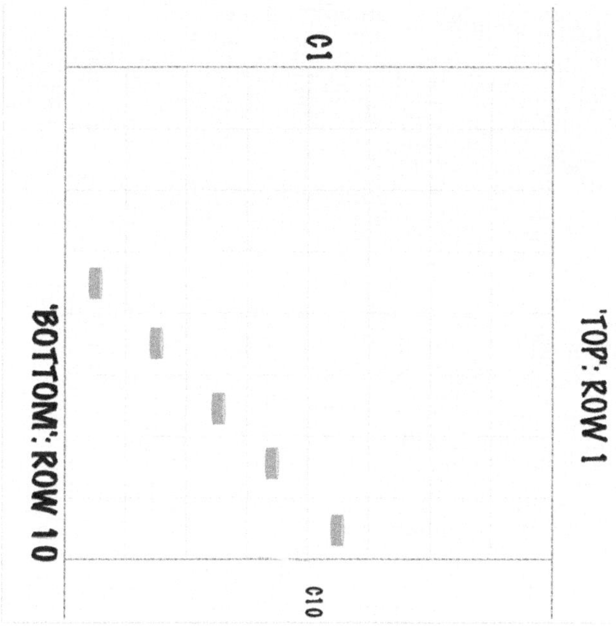

Recommended Placement of Pieces for Round Three:
One particle moving at constant velocity.

~

4

ROUND FOUR

Play is identical to that of *Round Three,* but for the following small adjustment:

- As Thedama will tell Aneyewit,
- For this round, the sample size has shifted again to *1 - 3* particles.

Recommendation for Thedama:

Select three particles:

Have two travel at a velocity of 1 square per square.

Have the third travel twice as fast.

Consider, for example:

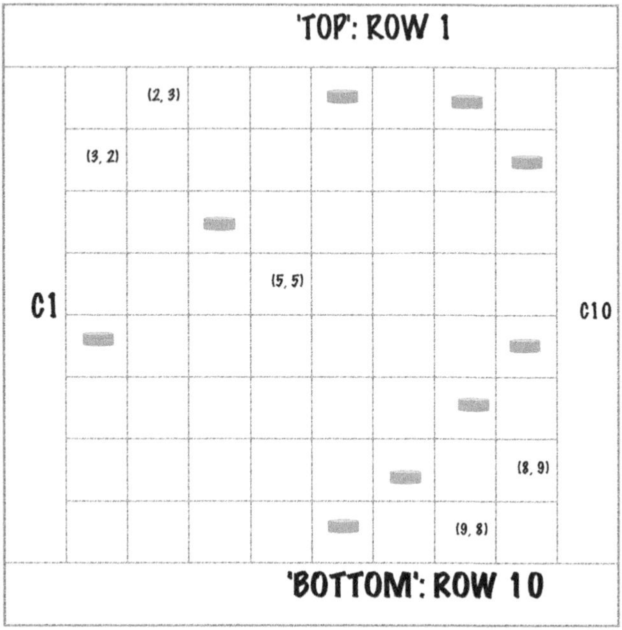
'TOP': ROW 1
(2, 3)
(3, 2)
(5, 5)
C1
C10
(8, 9)
(9, 8)
'BOTTOM': ROW 10

Made in the USA
Coppell, TX
05 November 2026

11286709R10049